# Contents

I. Background and Perspective .......................................................................... 1

II. Historical Technology Convergence ............................................................ 3

III. The Army's Future Capabilities—Warfighter Outcomes ............................. 5

IV. Technological Convergence: Human Dimension—Biobehavior and Warfighter
Resilience ...................................................................................................... 7

V. Technological Convergence—Virtual Presence, Enhanced Sensing, and Augmented
Autonomy ...................................................................................................... 13

VI. Technological Convergence: Scientific Underpinnings ............................. 20
    Mechanochemical Transduction ................................................................ 21
    Quantum Information Science ................................................................... 25

VII. The Army's Next Study—Comments and Recommendations .................... 28

VIII. Conclusion ............................................................................................. 31

Appendix: Expanding on Human Warfighter Outcomes—Biobehavioral Response ....... 32

# I. Background and Perspective

An important challenge for the Department of Defense (DOD) science and technology (S&T) programs is to avoid technological surprise resulting from the exponential increase in the pace of discovery and change in S&T worldwide. The nature of the military threat is also changing, resulting in new military requirements, some of which can be met by technology. Proper shaping of the S&T portfolio requires predicting and matching these two factors well into the future. Some examples of technologies which have radically affected the battlefield include the Global Positioning System coupled with inexpensive hand held receivers, the microprocessor revolution which has placed the power of the Internet and satellite communications into the hands of soldiers in the field, new sensing capabilities such as night vision, the use of unmanned vehicles, and composite materials for armor and armaments. Some of these new technologies came from military S&T, some from commercial developments and still others from a synthesis of the two sectors; but all were based on advances in the underlying sciences. Clearly, leaders and planners in military S&T must keep abreast of such developments and look ahead as best they can.

Since World War II, predictions of S&T for enabling military capabilities have occurred periodically. A study chartered by the Army Air Force[1] in 1947 predicted a broad range of developments in aeronautics and air power, and the study process has been a model for such forecasts ever since. Projections in S&T have been issued for many years by the National Research Council (NRC) of the National Academies, and the NRC occasionally publishes decadal studies for specific disciplines. NRC committee reports for astronomy and astrophysics, for example, go back every 10 years to at least 1964.

In DOD, the last series of forecast studies was done in the 1990s.[2] The Center for Technology and National Security Policy (CTNSP) in 2008 assessed the Army-sponsored Strategic Technologies for the Army of the Twenty-First Century (STAR 21) study[3] in which the basic and applied sciences were assessed and forecast as separate and discrete disciplines. Future capabilities were discussed in a separate set of STAR 21 volumes on systems and, in general, the technologies comprising individual systems were not

---

[1] The publication of *Toward New Horizons* by the new Army Air Force Science Advisory Group chaired by Theodore Von Karman charted the way ahead for air power for the United States. The history of this study is in: H. Gorn (editor), *Prophecy Fulfilled, 'Toward New Horizons' and its Legacy* (Washington, DC: Air Force Historical Studies Office, 1994). Theodore von Karman, *Toward New Horizons* (Washington, DC: United States Army Air Force, 1945). Available at <http://www.airforcehistory.hq.af.mil/Publications/authorindex htm>.

[2] Board on Army Science and Technology, Commission on Engineering and Technical Systems, and National Research Council, *STAR 21—Strategic Technologies for the Army of the Twenty-First Century* (Washington, DC: National Academy Press, 1992); Naval Studies Board and National Research Council, *Technology for the United States Navy and Marine Corps, 2000–2035* (Washington, DC: National Academy Press, 1997); Air Force Scientific Advisory Board, *New World Vistas, Air and Space Power for the 21ˢᵗ Century*, (Washington, DC: The Department of the Air Force, The Pentagon, 1995).

[3] John Lyons, Richard Chait, and Jordan Willcox, *An Assessment of the Science and Technology Predictions in the Army's STAR 21 Report,* Defense & Technology Paper 50 (Washington, DC: Center for Technology and National Security Policy, National Defense University, July 2008).

discussed with reference to the underlying sciences. This separation of future capabilities from the underlying S&T forecasts was similar for the studies of all three Services.

As forecasting is not an exact science, follow-on analyses of such efforts are necessary to determine the accuracy of initial claims. Therefore, CTNSP undertook an assessment of the initial STAR 21 report[3] to evaluate the original predictions 15 years later, in 2008. It was found that approximately one quarter of the predictions were on target, while others fell in a wide range of error (both over and underestimating the impact of particular advances), including some that proved completely inaccurate. The most egregious anticipatory errors were noted in failing to predict the impact of proliferation of the Internet, wireless technological capabilities, and personal computing devices. However, even considering these shortfalls, the STAR 21 study served a significant underlying purpose: educating the Army's leaders as to the nature and impact of S&T.

In the follow-on to the STAR 21 report, we recommended new approaches to future studies[4] and made suggestions as to how to proceed with their implementation. We recommended approaches to these studies, including spreading them out over several years and dividing them into focused portions of the sciences and technologies of interest each year. We also urged that contractors with maximum credibility, such as the NRC, perform the studies jointly with the other Services. The recommendation for tri-Service studies is still under discussion; therefore, the present proposal is limited to the Army.

Regarding approaches to new studies, we recommended technology convergence be emphasized. Here, the concept is two-fold: first, list the outcomes and associated capabilities desired by the Army at a future time point, and then look for confluences or convergences of individual sciences and technologies that would enable the realization of such capabilities. This would be achieved through forming clusters of sciences and technologies judged to be likely sources of such convergences and forecasting the evolution of these component topics on a common timeline. A resulting roadmap for each cluster would highlight where convergences of matured subjects were likely to occur and, therefore, where the capabilities and outcomes would be realized.

As the Army considers its next comprehensive technology forecasting effort as a follow-on to STAR 21, it should not only take into account the concepts of technology convergence as explained in Section II, but also tie them together with the capabilities desired by the warfighter. These are known as warfighter outcomes and are discussed in Section III. Sections IV–VI then present examples of technology forecasting, applying the concepts of convergence and warfighter outcomes to areas of interest to the Army.

---

[4]John W. Lyons, Richard Chait, and James J. Valdes, *Forecasting Science and Technology for the Department of Defense,* Defense & Technology Paper 71 (Washington, DC: Center for Technology and National Security Policy, National Defense University, December 2009).

# II. Historical Technology Convergence

Early convergences led to the invention of radar, which arose from the application of electromagnetic radiation (science) to how it interacts with materials, culminating in the development of microwave generators, transmitters, and power supplies among other devices (technology).[5] The technology so heavily employed today in wireless handheld devices coupled with Internet access dates back to technological advances in the science of telephony and advances in solid-state physics. These advances have provided processes for developing the fiber optics and computer technology enabling broadband network service throughout the world.

Convergences in S&T also occurred within the life sciences disciplines. Most notably this occurred early on in 1953 via the discovery of DNA's double helix structure by Watson and Crick.[6] A confluence of organic chemistry, physics, genomics, and information technology further provided the ability to amplify and replicate the DNA molecule in mass quantities, leading to advances in protein sequencing and synthesis in the 1980s. These initial convergences led to such advances as the sequencing of the human genome. Further innovations in information technology (IT) made possible a deeper understanding of the workings of human gene interactions, from which new fields of science have emerged such as genomics, proteomics, transcriptomics, and metabolomics. Modern medicine bases many of its practices on information resulting from this process of continuous convergences, making a strong case for an emphasis on forecasting further S&T convergences.

Forecasting S&T convergences requires care to avoid the stovepipe effect. It is imperative to assess capabilities derived from the conjunction of many systems. "Roadmapping" provides an effective method for tracking convergences and developments over time not only to detail historical convergences, but also, in turn, to effectively predict future convergences based on similar patterns.

---

[5] Timothy Coffey, Jill Dahlburg, and Elihu Zimet, *The S&T Innovation Conundrum*, Defense & Technology Paper 17 (Washington, DC: Center for Technology and National Security Policy, National Defense University, August 2005).

[6] James D. Watson and Francis Crick, "A Structure for Deoxyribose Nucleic Acid," *Nature,* 171 (1953), 737.

# III. The Army's Future Capabilities—Warfighter Outcomes

To complement the idea of S&T convergence and assure that forecasting is relevant to DOD needs, it is important to consider the desired future capabilities and warfighter outcomes. As noted above, we recommend that future studies focus on clusters of particular sciences or technologies. Each cluster would be coupled with one or more desired capabilities necessary to achieve warfighter outcomes defined by the U.S. Army Training and Doctrine Command (TRADOC). Warfighter outcomes of high level importance are:[7,8]

1. Counter Improvised Explosive Device (IED) and Mine Integrated Warfighter Outcome. The Future Force must have the ability to detect and neutralize chemical, biological, radiological, nuclear, and high yield explosives (CBRNE) obstacles and/or their components from a standoff distance of 100 meters.

2. Battle Command Network Integrated Warfighter Outcome. The Future Force must possess worldwide, beyond-line-of-sight network capabilities that are effective, layered, and protected.

3. Training Integrated Warfighter Outcome. Provide Soldiers and leaders the ability to excel in a challenging and increasingly complex future operating environment by developing tools and technologies that enable more efficient training through live, virtual, immersive and adaptable venues.

4. Power and Energy Integrated Warfighter Outcome. Provide enhanced agility to operate worldwide by reducing by half, the weight and volume of fuel associated with powering the force.

5. Human Dimension Integrated Warfighter Outcome. The Army leverages enhanced means to identify, access, retain, and develop Soldiers with unsurpassed cognitive, physical, and social (moral and cultural) capabilities. Soldiers are enabled by technology and combinations of cognitive, medical and social sciences to achieve excellence in small unit competence and to dominate in increasingly complex operational environments. Soldiers are able to leverage technologies and processes that optimize and restore cognitive and physical performance.

This paper offers several examples of clusters that could be the subject of study by an Army-chartered independent group. Each of the long-range forecasts would cover a period of from 15 to 25 years. To make the exercise manageable, we recommend that only a few clusters be started in any one year so that the entire forecasting effort would be completed over several cycles. Also presented are the capabilities and outcomes with which these example clusters are matched, the nature of the obstacles to be overcome and the likelihood of success.

---

[7] Department of the Army, *Military Operations Force Operating Capabilities*, TRADOC Pamphlet 525–66 (Fort Monroe, VA: Training and Doctrine Command, March 2008).
[8] COL R.C. Effinger, "Warfighter Challenges/Warfighter Outcomes," Presented at the Technology Planning Conference, Adelphi, MD, May 2010.

Finally, we present examples of underpinning sciences, which may support several outcomes and hence are somewhat less focused on specific outcomes. These examples provide areas of interest we consider worthy of investigation as they reflect the current trends of science and technological development.

We begin with some examples addressing the Human Dimension warfighter outcome and the accompanying technological convergences.

# IV. Technological Convergence: Human Dimension–Biobehavior and Warfighter Resilience

Paul Bartone and James J. Valdes

There is a growing recognition of the importance of human factors for national security in defense policy and planning circles. For example, as noted previously, a recent comprehensive analysis of warfighter outcomes led to the identification of the human dimension as one of the "Big 5" Integrated Warfighter Outcomes for attention in FY11. The human dimension is described as follows:

> The Army leverages enhanced means to identify, access, retain, and develop Soldiers with unsurpassed cognitive, physical, and social (moral and cultural) capabilities. Soldiers are enabled by technology, cognitive, medical and social sciences to achieve excellence in small unit competence and to dominate increasingly complex operational environments. Soldiers are able to leverage technologies and processes that optimize and restore cognitive and physical performance.[7]

We can summarize the human dimension capability described here as "warfighter resilience and adaptability," or the capability for soldiers and leaders to maintain a high level of physical health and performance even under increasingly complex and stressful operational environments. "Resilience" is shorthand for the ability to withstand operational demands and stressors without breaking down, and "adaptability" refers to the ability to adjust quickly to changing environments and circumstances. The two go hand-in-hand, and adaptable, resilient soldiers and leaders represent a critical capability for the future force. The absence of resilience and adaptability is expressed in a wide range of human dimension problems in the military, from degraded mental and physical performance to serious negative mental health outcomes including post-traumatic stress disorder (PTSD) and suicide. In addition to long-term personal disability, the former may, negatively affect decisionmaking processes that directly impact mission success or failure and contribute, for example, to friendly-fire incidents and civilian combat casualties.

The top leaders of the Army have focused considerable attention on trying to prevent PTSD and soldier suicides. For example, the Army Vice Chief of Staff has issued numerous press releases throughout 2009 voicing the Army's concerns, with the goal of "trying every remedy and seeking help from outside agencies that are experts."[9] These concerns, and solutions to goals set by senior leaders, relate directly to research within the Human Dimension outcome.

TRADOC and senior military leaders (e.g., Secretary of Defense, Chairman of the Joint Chiefs of Staff, Army Chief of Staff and Deputy Chief of Staff, and Deputy Chief of Staff for Personnel) have identified adaptable military personnel as a critical capability for the

---

[9] U.S. Department of Defense, "Army Releases May Suicide Data," U.S. Department of Defense, Available at <http://www.defense.gov/Releases/Release.aspx?ReleaseID=12740>.

future.[10] In what follows, we briefly consider emerging trends and convergences in S&T that are likely to dramatically shift how we go about selecting, training and developing highly resilient and adaptable military personnel.

As previously noted, S&T convergence forecasting requires a focus on key disciplines that will lead to future developments capable of aiding the military. Three broad functional areas of selection, training, and treatment are central in achieving the warfighter resilience and adaptability capabilities sought. As with any organization, the start point for building a capable workforce is with the selection process, selecting in those with the needed qualities and capacities, and selecting out those who lack the needed qualities, or who possess undesired qualities. Next is training, which includes everything that the organization does to train and develop its workforce. Finally, treatment refers broadly to all programs and systems that aim to restore optimal readiness and performance following exposure to potentially damaging operational stressors. Research, diagnostic tools and instruments, and clinical trials are all important enablers that make it possible to improve selection, training and treatment. Figure 1 provides a representation of key scientific disciplines and technological themes that will impact the human dimension capability—warfighter resilience and adaptability— into the future.

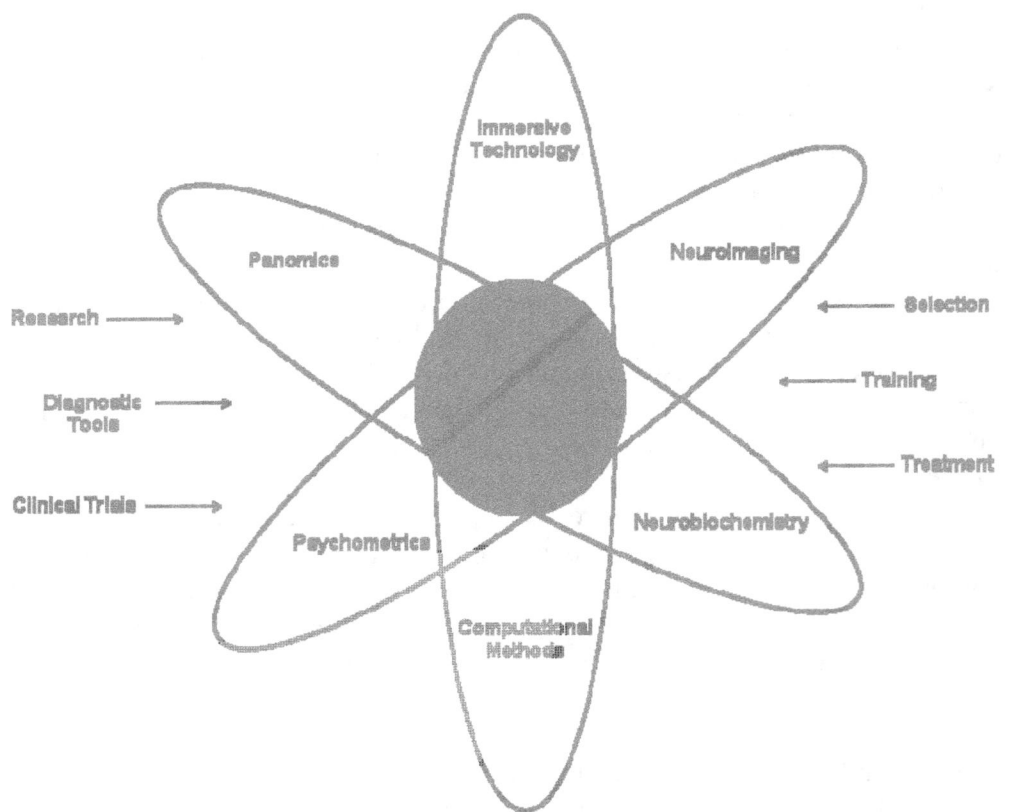

**Figure 1. Key Disciplines influencing the Human Dimension**

---

[10] Robert M. Gates, "The National Defense Strategy: Striking the right balance," *Joint Force Quarterly*, 52 (2009), 2–7.

As a more specific example of anticipated convergences in the Human Dimension area, Figure 2 displays how various S&T areas are interacting to create new knowledge and capabilities for a highly resilient and healthy defense workforce.

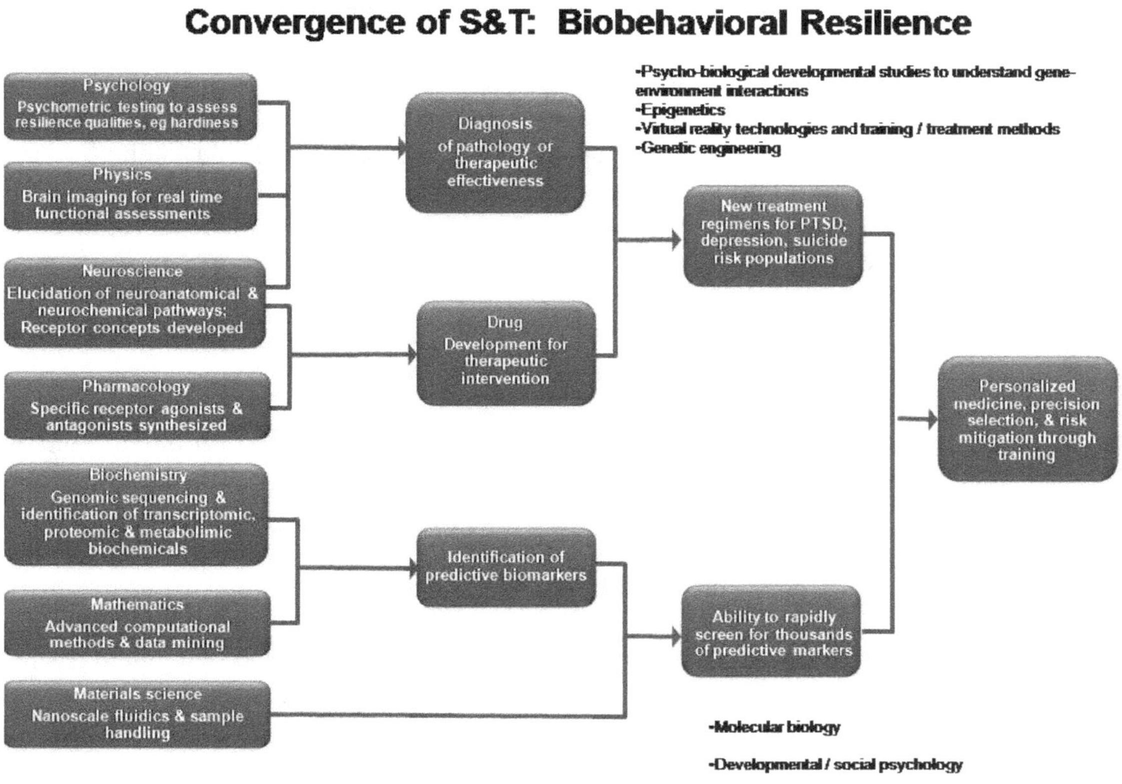

# Convergence of S&T: Biobehavioral Resilience

Figure 2. Technology Convergence for the Human Dimension—Biobehavioral Resilience

This figure displays how the parallel development of multiple disciplines and their likely convergences will lead to new and vastly more effective approaches to medical treatment and prevention that can be summarized as "personalized medicine", as well as more precise and effective selection and training programs in the human dimension. The field of neuroscience has been elucidating the neuroanatomical and neurochemical substrates of the brain. This knowledge, when combined with imaging techniques derived from physics, has resulted in the development of brain imaging techniques that allow a functional assessment of the brain. When coupled with psychometric testing predictive of resilience and adaptation, determinations of pathology or of therapeutic effectiveness can be made. Following the elucidation of specific receptors associated with particular brain substrates, advances in pharmacology led to the development of drugs with specific agonistic or antagonistic effects on these receptors, hence predictable effects on neural function and behavior. The new diagnostic tools and drugs could then combine to yield tailored treatment regimens for a range of neurological and psychological syndromes including depression and PTSD.

9

Meanwhile, the traditional field of biochemistry has branched into the "panomics", including genomics, transcriptomics, proteomics and metabolomics, all being extraordinarily information intensive. The parallel development of mathematical tools to mine and interpret these enormous data sets, and the advances in materials science, particularly nanoscale fluidics, has led to the design of high throughput sample analysis and the identification of predictive biomarkers for an enormous range of clinical conditions. The combination of this ability to have a comprehensive, real-time assessment of panomics and treatment regimens based on defined physiological substrates will result in the new field of personal medicine with the potential for personnel selection, risk mitigation through training or preventive medicine, and tailored therapies as needed. These same tools and technologies have also made possible new studies clarifying how genes and environmental factors interact to determine mental illness,[11] susceptibility to stress,[12] and social behavior.[13]

**Post-traumatic Stress Disorder and Suicide Prevention**

To date, studies of resilience, and risk or vulnerability to stress-related problems such as PTSD have focused either on psychological qualities or neurobiological factors independently. For example, Charney[14] summarizes evidence pointing to various neurotransmitters, neuropeptides, and hormones that appear to be linked to the psychobiological stress response and health outcomes, both good and bad. These include, but are not limited to, epinephrine, norepinephrine, neuropeptide Y, galanin, dopamine, serotonin, testosterone and estrogen—some of the very same markers biochemical assays can now efficiently and cost-effectively extract from very small tissue samples.

Advances in the field of psychology and psychometrics in recent years have led to better conceptual understanding of individual differences, as well as more reliable and valid measurement tools for assessing basic psychological qualities in people.[15,16] To take one relevant example, psychological hardiness, composed of commitment, control and challenge, has been identified in many studies as a consistent quality of individuals who display resilience under stress.[17]

Studies of resilience from a psychological standpoint have considered multiple variables, but generally have not attempted to assess what is going on at a biochemical level. To

[11] A. Caspi, K. Sugden, T.E. Moffitt, A. Taylor, I.W. Craig, H. Harrington et. al, "Influence of Life Stress on Depression: Moderation by a Polymorphism in the 5-HTT Gene," *Science*, 301 (2003), 386–389.
[12] F.A. Champagne and M.J. Meany, "Stress during Gestation Alters Postpartum Maternal Care and the Development of the Offspring in a Rodent Model," *Biological Psychiatry*, 59 (2006), 1227–1235.
[13] F.A. Champagne, "Genes in Context: Gene-Environment Interplay and the Origins of Individual Differences in Behavior," *Current Directions in Psychological Science*, 18 (2009), 127–131.
[14] D.S. Charney, "Psychobiological Mechanisms of Resilience and Vulnerability: Implications for successful adaptation to Extreme Stress," *American Journal of Psychiatry*, 161 (2004), 195–216.
[15] P.T. Costa and R.R. McCrae, *NEO PI-R. Professional manual.* (Odessa, FL: Psychological Assessment Resources, Inc., 1992).
[16] Timothy Judge and Joyce Bono, "Five-factor Model of Personality and Transformational Leadership," *Journal of Applied Psychology*, 85 (2000), 751–765.
[17] S.C. Kobasa, S.R. Maddi, and S. Kahn, "Hardiness and Health: A Prospective Study," *Journal of Personality & Social Psychology*, 42 (1982), 168–177.

continue with the example of psychological hardiness, studies with a variety of occupational groups have found that hardiness operates as a significant moderator or buffer of stress when the outcome is some measure of health or illness.[11,18] In addition, some studies have shown that low hardiness is associated with negative or destructive coping behaviors (e.g., substance abuse) under stressful situations.[19] Hardiness has also been identified as a moderator of combat exposure stress in Gulf War soldiers[20] and emerged as a stress buffer in other military groups as well, including U.S. Army casualty assistance workers,[21] peacekeeping soldiers,[22] Israeli soldiers in combat training,[23] Israeli officer candidates, and Norwegian Navy cadets.[24] Further, studies have found that troops who develop PTSD symptoms following exposure to combat stressors are significantly lower in hardiness compared to those who do not develop PTSD.[25] Under low-stress conditions, troops high in hardiness report about the same level of PTSD symptoms as those low in hardiness. However, under high-stress conditions, those high in hardiness report significantly fewer PTSD symptoms than those low in mental hardiness. Other studies have found similar stress and hardiness interaction effects among combat-exposed veterans. These results suggest that those who are high in the qualities of hardiness are more resistant to the ill effects of operational stress. But the underlying biological and physiological processes associated with this type of stress resilience are just starting to be systematically addressed.

For example, a recent study conducted at NDU found that psychological hardiness is associated with significantly higher levels of high-density lipoprotein (HDL).[26] It is known that higher levels of HDL are protective against coronary heart disease, in part by more efficiently disposing of excess lipids that can contribute to a range of cardiovascular disorders.[27,28] It may be that the kinds of positive coping expectations and behaviors that

[18] R.J. Contrada, "Type A Behavior, Personality Hardiness, and Cardiovascular Responses to Stress," *Journal of Personality and Social Psychology*, 57 (1989), 895–903.

[19] S.R. Maddi, P. Wadhwa, and R.J. Haier, "Relationship of Hardiness to Alcohol and Drug Use in Adolescents," *American Journal of Drug and Alcohol Abuse*, 22 (1996), 247–257.

[20] P.T. Bartone, "Hardiness as a Resiliency Factor for United States Forces in the Gulf War," in *Posttraumatic Stress Intervention: Challenges, Issues, and Perspectives*, edited by J.M. Violanti, D. Paton, and C. Dunning, 115–133. Thomas: Springfield, IL, 2000.

[21] P.T. Bartone, R.J. Ursano, K.M. Wright and L.H. Ingraham, "The Impact of a Military Air Disaster on the Health of Assistance Workers: A Prospective Study," *Journal of Nervous and Mental Disease*, 177 (1989), 317–328.

[22] T.W. Britt, A.B. Adler and P.T. Bartone, "Deriving Benefits from stressful Events: The Role of Engagement in Meaningful Work and Hardiness," *Journal of Occupational Health Psychology*, 6 (2001), 53–63.

[23] V. Florian, M. Mikulincer and O. Taubman, "Does Hardiness Contribute to Mental Health During a Stressful Real Life Situation? The Role of Appraisal and Coping," *Journal of Personality and Social Psychology*, 68 (1995), 687–695.

[24] P.T. Bartone, B.H. Johnsen, J. Eid, W. Brun, and J.C. Laberg, "Factors Influencing Small Unit Cohesion in Norwegian Navy Officer Cadets," *Military Psychology*, 14 (2002), 1–22.

[25] P.T. Bartone, "Hardiness Protects Against War-Related Stress in Army Reserve Forces," *Consulting Psychology Journal*, 51 (1999), 72–82.

[26] Paul T. Bartone, Tony Spinosa and Joel Robb, "Psychological Hardiness is Related to Baseline High-Density Lipoprotein (HDL) Cholesterol Levels," (presentation, annual convention of the Association for Psychological Science, San Francisco, May 24 2009).

[27] P. Barter, "The role of HDL cholesterol in preventing atherosclerotic disease," *European Heart Journal*, Supplement 7 (2005), F4–F8.

mark high hardy persons have a direct effect on HDL production by moderating the excretion rate of stress hormones that also influence cholesterol production, which serves to protect against cardiovascular disease.[29] These identified biochemical correlates of hardiness are also suggestive of an underlying genomic basis for resilience, which might be assessed via the proteins for which these genes code.

Although such studies indicate that hardiness has identifiable biological correlates, most of the potential biomarker candidates have not yet been examined. Beyond cardiovascular and basic blood chemistry (e.g., cholesterol) markers associated with stress response, a growing body of research points to a number of neurochemical substances and processes related to response to stress and healthy or unhealthy outcomes. (Further information on biochemical and neurological correlates is given in the appendix.)

Advances in technologies for assessing both psychological qualities and neurochemical processes are leading toward convergence across these domains, which should result in new cross-disciplinary studies and more complete understanding of the stress resilience and adaptation process. These advances will in turn make possible more precise and effective strategies for treatment, as well as for training and selection.

However, the range of S&T convergences within the human dimension extends far beyond genetic and biological spheres. Warfighter outcomes can also be achieved through the enhancement of soldiers' situational awareness through a combination of physical, mechanical, and materials sciences and technologies. These are brought to light through such advancements as virtual presence and enhanced sensing.

---

[28] P.M. Ridker, M.J. Stampfer and N. Rifai, "Novel Risk Factors for Systemic Atherosclerosis," *JAMA*, 285 (2001), 2481–2485.
[29] R. Fraser et al, "Cortisol Effects on Body Mass, Blood Pressure, and Cholesterol in the General Population," *Hypertension*, 33 (1999), 1364–1368.

# V. Technological Convergence—Virtual Presence, Enhanced Sensing, and Augmented Autonomy

Albert Sciarretta

It should be noted that in addition to 6.1 (basic research), convergence includes the tiered fusion of technologies within the 6.2 (applied research) or 6.3 (advanced technology development) levels of S&T. For example, microelectromechanical system (MEMS) technology is the integration of mechanical, sensor, electronics, software, and materials science technologies. These technologies provide the sensing, processing, and actuator functions as well as the platform/case for the MEMS. Then, inertial measurement unit (IMU) technology may be a combination of MEMS accelerometers, MEMS gyroscopes (for measuring changes in rotational attributes such as roll, pitch, and yaw), and additional sensors (via an intelligent fusion algorithm) used to reset the IMU (to reduce drift). At the next higher level, 3-D location/tracking technology for a ground robotic system may be a combination of MEMS-based Global Positioning System (GPS) sensors, IMUs, and an intelligent fusion algorithm (e.g., includes some intelligent processing in addition to a Kalman Filter). The intelligent fusion algorithm may select the GPS sensors as the primary data source when out in the open, the IMUs while inside buildings, and a combination when transitioning from outdoors to indoors and vice versa. The intelligent fusion algorithm may also provide feedback to reset an IMU. This can be depicted as follows:

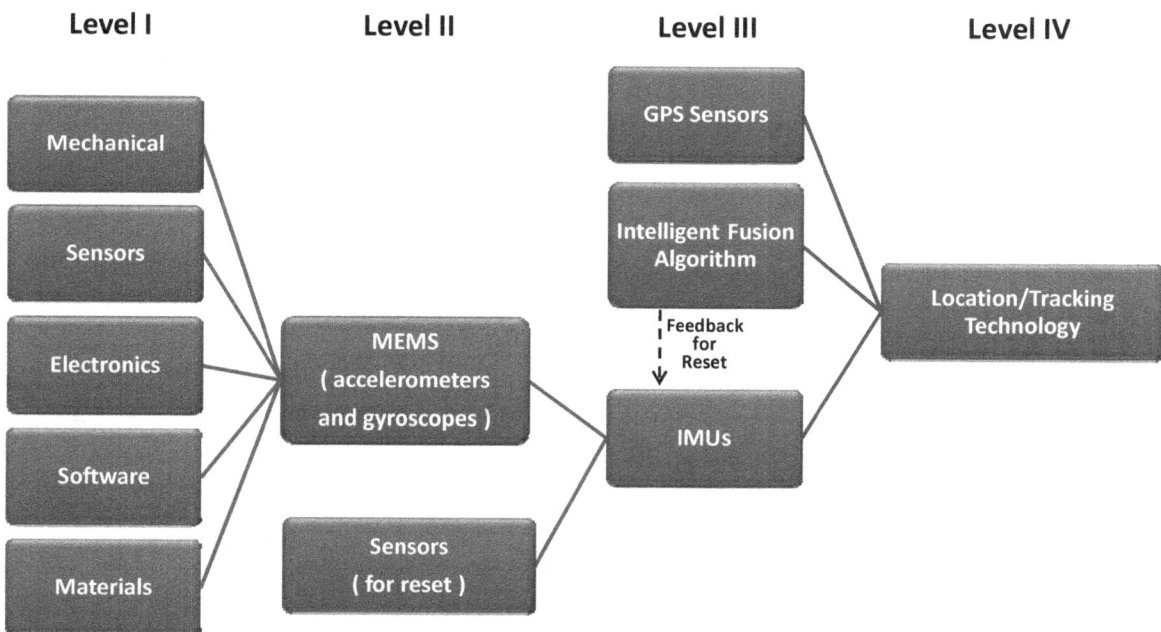

**Figure 3. Roadmap of S&T Convergence in Materiel—Location and Tracking Technology**

One should note that sensor technology is depicted twice in the Level I technologies and once each in the Level II and Level III technologies. This redundant occurrence of a technology, may be an indication that it is a critical technology, and also that its development will support broad applications. It is also an indication that there is most likely a convergence of technologies (electro-optical, magnetic, acoustic, electronics, materials, etc.) within the sensor technology area.

The above example provides a "bottom up" approach for describing convergence. To continue on to a Level V technology, say "robotics," may prove to be a difficult task (e.g., identify other Level IV technologies converging with "location/tracking technology.") In addition, there may be one or more intermediate levels between Level IV technology and the targeted "robotics" technology level. Thus, a better approach may be to use a "top down" or "decomposition" analysis to identify the various levels of converging technologies.

## Virtual Presence and Enhanced Sensing Capabilities

For the human dimension integrated warfighter outcome, the Army should leverage enhanced means to identify, access, retain, and develop soldiers with unsurpassed cognitive, physical, and social (moral and cultural) capabilities. Soldiers should be enabled by technology, cognitive, medical, and social sciences to achieve excellence in small unit competence and to dominate in increasingly complex operational environments. Soldiers should be able to take advantage of technologies and processes that optimize and restore cognitive and physical performance.

To provide these characteristics, disparate S&T efforts might converge to provide two capabilities that enhance a warfighter's situational awareness and a third capability that will enhance the autonomy of robotic systems. The two capabilities that enhance a warfighter's situational awareness are *virtual presence* and *enhanced sensing capabilities*. These capabilities are based on the wireless linkage of a warfighter's mind to the intelligence and sensing capabilities of one or more robotic systems.

Virtual presence for a warfighter is the ability of a warfighter to experience critical information in the context of an environment/situation that is not immediately surrounding the warfighter. More specifically, for this paper, it is the full immersion of the warfighter's mind into the environment/situation being experienced by a robotic system that is not in the warfighter's line of sight. A warfighter will have the sensation that he can virtually be in the same room as and in the same position and orientation of the robotic system. This virtual presence would provide a vivid feeling of being detached from one's body; so much so, that if the robotic system were to position itself to look back at the warfighter, the warfighter would have the sensation of an "out of body" experience.

As opposed to a mental detachment from one's body, enhanced sensing capabilities are an expansion of the warfighter's physical sensing capabilities. For example, a warfighter will sense (at least see and hear, perhaps smell and feel) what a remote, non line-of-sight robot system (e.g., in another room of a building) is sensing. Seeing something in the "corner of one's eye" will have new meaning, in that the warfighter will not only sense his immediate surroundings but also the surroundings of the robotic system. It is unclear, at this point, how these enhanced sensing capabilities will be interpreted by the warfighter. Perhaps, for sight, it will be similar to a screen within a screen. Perhaps for hearing, it will be similar to "Superman-like" hearing.

## *Augmented Autonomy*

The third capability created by the convergence of science and technology is the enhancement of the autonomy of a robotic system—*augmented autonomy*. This convergence primarily benefits a robotic system, but indirectly benefits the two warfighter enhancements above—by making robotic systems more intelligent. It may be many decades before the development of fully autonomous robotic systems, especially those that can learn and adapt their behaviors to changing situations. Regardless, there are many technical challenges that will hamper researchers from developing fully autonomous robotic systems in the near future.

To make up for shortcomings in reaching full autonomy, a robotic system's "brain" might be connected to its human controller's brain, so that some of the robotic system's "thinking" is being accomplished by the human. Thus, the robotic system's artificial intelligence (AI) is made up of a combination of the AI programmed into its processor and the intelligence of the human mind connected to the robotic system—a sort of "distributed intelligence" capability. The robotic system's autonomy (AI) is augmented by its human partner's brain. This augmented autonomy may simply assist the robotic system in better understanding which direction to move next; or it may assist the robotic system in making sense of complex, ambiguous information, such as the military status of a child aiming a toy gun at the robotic system.

For this paper, convergence of S&T is being considered in a similar manner as previously described in the beginning of this section for location and tracking technologies. Figure 4 is a depiction of S&T that might converge to provide the three capabilities described above. In addition to the convergence of the S&T, Figure 4 also depicts how the enhancement of the behavior of robotic systems—through augmented autonomy—will improve autonomous robotic behavior, and thus the capabilities of semi-autonomous and autonomous robotic systems. This enhancement will, in turn, further improve the two warfighter enhancements.

The development of *virtual* presence technologies will most likely depend on advances in four major technology areas: gaming technologies, immersive technologies, semi-autonomous and autonomous robotic systems, and mind-machine interfaces. *Enhanced sensing* capabilities will depend primarily on the latter three, while *augmented autonomy* will benefit from advances in the latter two. The technologies contributing to S&T convergences (far left of figure 4) converging to these four major technology areas are described below.

15

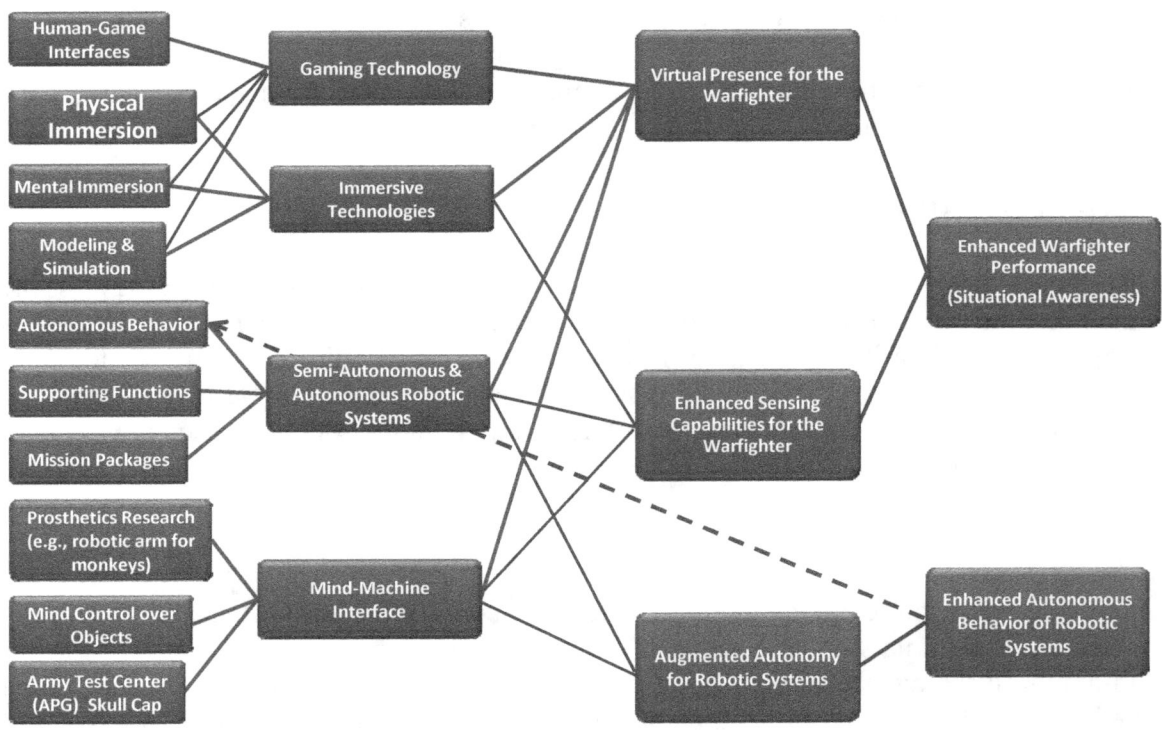

**Figure 4. S&T convergence (from left to right) to enhance warfighter situational awareness and the autonomous behavior of robotic systems**

*Human-game interfaces* can provide insights for both physical and mental immersion of warfighters with the robotic systems. For example, the most popular games very successfully immerse the cognitive processes of the player with the game itself —especially those associated with massive multi-player online games using virtual worlds. Gamers also have access to physically immersive tools (e.g., tactile vests that provide thuds when kicked in a kick boxing game, helmet mounted displays, data gloves that provide positive feedback when one grabs something in a virtual world).

*Physical immersion* uses technologies that immerse a human into a virtual, yet almost realistic, environment. These technologies exploit recent breakthroughs in:

- Miniaturization/advancement of immersive multi-sensory, multi-mode technologies (gesture/body motion recognition, occlusion visualization)
- Advances in annotation filters, view management, context models, and learning/reasoning agents

Work on physical immersive systems is being pursued by the Army's Human Research and Engineering Directorate (HRED) Field Office at Fort Benning, GA. For example, HRED has experimented with a tactile land navigation system.

*Mental immersion* uses technologies that stimulate a brain to believing it is in a realistic environment. Dreams are a good example of mental immersion. Though not as realistic in sensation, on-line games may stimulate players so much that the players believe they have a "second life" within the game—they are immersed in an alternative world.

*Modeling and simulation*, with the use of physical immersion technologies, can contribute greatly to the creation and rendering of virtual worlds. This area indirectly supports the functional capabilities of a robotic system and supports its development along with systems engineering.

*Autonomous behavior, supporting functions,* and *missions packages* are components of unmanned and autonomous systems. The subcomponents of each are listed below.

- *Autonomous behavior* can be broken down into:
  o Perception—a robot's ability to sense and observe its surroundings by relating features in sensor data to those in the real world
  o Planning—(path and mission) Path planning, deliberate or reactive, includes the development of a movement trajectory from current to next position(s). Mission planning provides the best course of military action; given situational awareness of environment, enemy situation, friendly situation, weather, time available, etc.; assessment of assigned mission order; doctrine and tactics; standard operating procedures; and other related military planning information.
  o Navigation—situational awareness of the movement space, knowing current location and direction of movement, knowing directions to desired location(s), ability to map and find a way through immediate surroundings, and ability to detect nearby hazards to mobility
  o Behavior and Skills—the combination of AI with inputs from perception, planning, and navigation to support cooperative behavior and develop motor commands. These motor commands may include mobility commands for flying, walking, jumping, etc.; military-related commands for communicating information, maneuvering with other forces, firing weapons, accomplishing assigned missions (e.g., surveillance, reconnaissance, explosive ordnance detection, etc.); commands for interacting with human controllers (including commanders), other humans (friendly military, enemy, and non-combatants), and other robotic systems; control of mission packages (see below); and for the accomplishment of related skills (e.g., control actuators to pick up a Styrofoam cup of coffee without crushing it, and later control the same actuators to crush a weapon to make it inoperable).
  o Learning and Adaptation—the ability of a robotic system to enhance or modify its artificial intelligence as well as its behavior. For example, the ability of a robotic system to learn new military roles when transitioning from conventional to irregular warfare, including learning new rules of engagement and adapting to changing enemy tactics.
- *Supporting functions* include:
  o Mobility—ability of a robotic system to traverse through space, air, land (natural and manmade terrain), and water (on and under) environments; including combinations of those environments (e.g., systems that operate in

littoral regions; systems that operate on the ground but also fly to traverse obstacles or large distances).

- o Robot-Human, Robot-Robot, and Robot-C4ISR[30] System Interactions— ability to interact with other humans, robots, and C4ISR systems, which includes teamwork (e.g., organize into teams and allocate tasks); ability to communicate with others; and ability to understand "commander's intent."
- o Communications—ability to convey and receive concepts with voice, communications systems including military and non-military communications links), nonverbal communications (e.g., hand-and-arm signals, gestures), and graphical user interfaces.
- o Power and Energy— includes power sources (e.g., rechargeable and non-rechargeable batteries, fuel cells, engines) and energy management to support the robotic system and its mission package.
- o Health Maintenance— ability to make the robotic system more robust and to provide maintenance capabilities for self-monitoring, diagnostics, and recovering from component failures.
- *Mission packages* include:
  - o Modular physical components—(e.g., lethal weapon, non-lethal weapon, surveillance system) that are attached to a common robotic platform to provide it a unique capability
  - o Modular software components—(e.g., complex "decision-focused" software, simplistic "decision focused" software, less complex "follow orders" software) to delineate higher echelon "leader" robotic systems, lower echelon leaders, "follower" robotic systems, or even one-of-a-kind task (e.g., robotic sensor) robotic systems

*Prosthetics research* has made great strides because of the miniaturization of supporting technologies and power systems, as well as enhancements in processing and control. Increased DOD interest has also provided needed focus in this area. Of particular interest for this convergence is the linking of the brain to a robotic arm.[31] A current example of success in this area is the demonstration of a monkey mentally controlling a robotic arm to feed itself a banana. This mental control results from a brain-machine interface in which tiny electrodes are implanted into the motor cortex of the monkey's brain. A 61% task completion rate was noted in a mere two days of training, during which the monkey's neuronal signals are transferred into directional signals for the robotic arm. It is assumed that similar mind-over-robot control can be developed with wireless connections.

*Mind control over objects* has been demonstrated in the form of a game with a wireless headset. Electrical activity in the brain translates to signals understood by a computer

---

[30] Command, control, communications, computers, intelligence, surveillance, and reconnaissance
[31] "Monkeys control robots with their minds," CNN, available at <http://www.cnn.com/2008/TECH/science/05/29/monkey robots/index html>.

controlling a fan that blows a small ball up into the air.[32] The harder an individual concentrates, the more electrical activity generated, translating to faster propulsion of the fan which blows the ball higher. Obviously, this is very primitive control. However, given the advancements in prosthetics research and this demonstration, wireless mind-machine interfaces seem achievable.

Interestingly, the U.S. *Army Aberdeen Test Center* (the ATC at the Aberdeen Proving Ground) has demonstrated the ability to tap into the workings of a warfighter's brain in an operational test environment.[33] The ATC's skull cap utilizes advances in electric potential sensors to detect bioelectric signals (electro-encephalograms [EEG] and electro-cardiograms [EKG]) in atypical environments (e.g., through clothing and hair, in the presence of water-based fluids like sweat) without requiring skin contact or conductive gels. Advanced algorithms process the signals to estimate the warfighter's cognitive workload, engagement, and fatigue levels.

---

[32] Joel Garreau. "What If You Could Move Objects With Your Mind? Well, That Time Has Come," Washington Post Report, available at <http://www.washingtonpost.com/wpdyn/content/article/2009/04/22/AR2009042204036 html>.

[33] James Buxton, "Aberdeen Test Center Warfighter Assessment," (provided by email to the author, Aberdeen Test Center, March 1, 2010).

# VI. Technological Convergence: Scientific Underpinnings

Prior to implementing such efforts, however, it is important to remain aware of and to understand the scientific underpinnings that could provide the foundation for the technology convergence discussed above. Two examples of scientific underpinnings worthy of note are mechanochemical transduction and quantum information science, both areas that suggest convergences of chemistry, physics, mathematics, biology, and materials science.

# Mechanochemical Transduction

Douglas Kiserow and Kelby Kizer

The applications of heat, pressure, or electrical energy are typical approaches used for activating and controlling chemical reactions. A chemist uses these energies to activate chemical reactions, thereby providing the common methods used to generate polymers, ceramics, and other modern materials.[34,35] However, once materials are created they are susceptible to damage from other forces, such as the energy released from a physical impact. Despite the variety of techniques available to chemists, the response of materials to mechanical energy (e.g., physical damage) is the undesired breaking of chemical bonds, resulting in cracking and structural failure.[34,36,37,38]

A revolutionary step in engineering would be possible if materials could be constructed such that the energy released from mechanical damage could be harnessed for useful chemical reactions, such as rebuilding broken bonds.[34] Surprisingly, a student of Aristotle over 2,300 years ago made the first recorded empirical observation that mechanical energy applied with a mortar and pestle to certain compounds produced a unique chemical change.[34,39] Although it was not understood at the time, chemists later identified this as the first recorded use of mechanical energy (in this case, grinding) to activate a chemical reaction.[34] In addition to simple grinding, mechanical energy can include the relatively modest forces released by scratching, hitting, and shaking, as well as the more dramatic forces released by earthquakes, explosions, and projectile impacts. Polymer chemists hypothesized as early as the 1950s that mechanical energy provided by damaging forces such as a projectile impact could be harnessed to provide the necessary energy to activate predefined chemical reactions of choice, rather than causing the undesired breaking of carbon-carbon bonds in a polymer backbone.[34,40] Unfortunately, this initial hypothesis could not be tested effectively until the past decade.

The idea of harnessing the use of mechanical energy to activate chemical reactions, termed "mechanochemistry" or "mechanochemical transduction," has come to the forefront of research efforts in chemical and materials science.[34,41] The recent cross-disciplinary bridging of research efforts in materials science and chemistry have been essential in bringing this field to life, as is clear when one considers the following summary of the fundamental discoveries in mechanochemistry described here.

---

[34] M.K. Beyer and H. Clausen-Schaumann, "Mechanochemistry: The Mechanical Activation of Covalent Bonds," *Chemical Reviews,* 105 (2005), 2921–2948.

[35] B.M. Rosen and V. Percec, "Mechanochemistry: A Reaction to Stress," *Nature, 446* (2007), 381–382.

[36] C.R Hickenboth, J.S. Moore, S.R. White, N.R. Sottos, J. Baudry, S.R. Wilson, "Biasing Reaction Pathways with Mechanical Force," *Nature,* 446 (2007), 423–427.

[37] C. Weder, "Mechanochemistry: Polymers React to Stress," *Nature,* 459 (2009), 45–46.

[38] V.V. Boldyrev, K. Tkácová, "Mechanochemistry of Solids: Past, Present, and Prospects," *Journal of Materials Synthesis and Processing,* 8 (2000), 1064–7562.

[39] H. Staudinger and H.F. Bondy, "Über Isopren und Kautschuk," *Berichte der Deutschen Chemischen Gesellschaft,* 63 (1930), 734–736.

[40] W.F. Watson, "Die Makromolekulare Chemie," *Makromoekulare Chemie,* 34 (1959), 240–252.

[41] M.M. Caruso, D.A. Davis, Q. Shen, S.A. Odom, N.R. Sottos, S.R White, and J.S. Moore, "Mechanically-Induced Chemical Changes in Polymeric Materials," *Chemical Reviews,* 109, (2009), 5755–5798.

An early piece of the mechanochemistry puzzle was reported by chemists in 2001, when a multi-ringed compound (spiropyran) was found to exhibit a color change from yellow to blue when mechanical force (again, grinding) was applied.[34] This color change was due to the breaking of a carbon-oxygen bond within the compound. However, this discovery involved an isolated compound that could be probed in powdered form within a mortar and pestle, and was a long way from demonstrating that mechanical force can be harnessed to activate specific chemical reactions in a bulk material. A further limitation of this initial finding was that the chemical change could also be due in part to heat released during grinding (caused by friction), rather than directly through mechanical force. It was only through the convergence of research efforts in chemistry and materials science that the discovery, synthesis, and necessary modifications of novel chemical species could be combined with the expertise required to fabricate and characterize responsive materials.[34,36,41] This confluence of disciplines directly led to the development of the first mechanically-responsive solid material and the multitude of new research efforts that followed.

A study led at the University of Illinois, including leading chemists and materials scientists at a variety of universities, has been bridging research in materials and chemicals sciences to better understand mechanochemical transduction.[36,41] In 2007, these investigators devised a coordinated approach for predicting, assembling, and testing new mechanochemical reaction pathways.[35,36] Chemists from the research group successfully simulated a new reaction pathway, predicting a mechanism whereby mechanical energy would activate a reaction and generate a chemical product that neither heat nor light could produce. Chemists then synthesized the compounds and incorporated the chemicals into a monomer. Materials scientists then took a lead role by characterizing the sample and found that mechanical energy (and not heat) activated the covalent bonds of the reactants, as predicted from the simulations.[35,36] Using this newfound knowledge, chemists then designed and synthesized mechanochemically active compounds, termed "mechanophores," and incorporated these compounds into solid, polymeric materials.

The first mechanophore to be tested in polymeric "bulk" materials was derived from spiropyran, the compound previously found to display a color change after mechanical force was applied.[34,36] This solid-state bulk material was assessed in 2009 using the characterization methods common in materials science. The results revealed that the solid material displayed mechanochemical activity. When a mechanical force (stretching) was applied to the solid, releasing energy, a chemical change in the spiropyran mechanophore resulted in a visible color change. The damage-sensing polymers provided the first proof-of-concept that solid materials could indeed be designed to harness energy from mechanical damage to activate preprogrammed chemical reactions.[37,42] Based on this revolutionary discovery, chemists and material scientists could design these reactions to convert mechanical damage to anything from self-repairing the materials to reconnecting broken circuits.

---

[42] D.A. Davis, A. Hamilton, J. Yang, L.D. Cremar, D. Van Gough, S.L. Potisek, M.T. Ong, P.V. Braun, T.J. Martínez, S.R. White, J.S. Moore, and N.R. Sottos, "Force-Induced Activation of Covalent Bonds in Mechanoresponsive Polymeric Materials," *Nature,* 459 (2009): 68–72.

Following these important observations, the team of chemical and materials science researchers began synthesizing a wide variety of mechanophores predesigned to exhibit a verity of responses, including self-sensing and self-healing activities.[43] A team of chemists and materials scientists led by the Moore group recently published results that demonstrated a new mechanophore released cyanoacrylate, the active ingredient of Super Glue®, when it was subjected to mechanical stress. Future research by this group of multidisciplinary investigators will involve the incorporation of this and other new mechanophores into polymeric materials, followed by evaluations of the corresponding responses to mechanical stress. If researchers can demonstrate that the new mechanophore-containing polymers respond to damage as designed, they will then investigate how the chemical structures respond when integrated into complex materials, such as polymer composites, metals, and/or ceramics.[41,43] These mechanically-responsive polymers may be the first stage in the development of a new field bridging the chemical and material sciences. The entire mapping of such possible technology convergences is shown in figure 5.

## Convergence of S&T: Mechanochemical Transduction

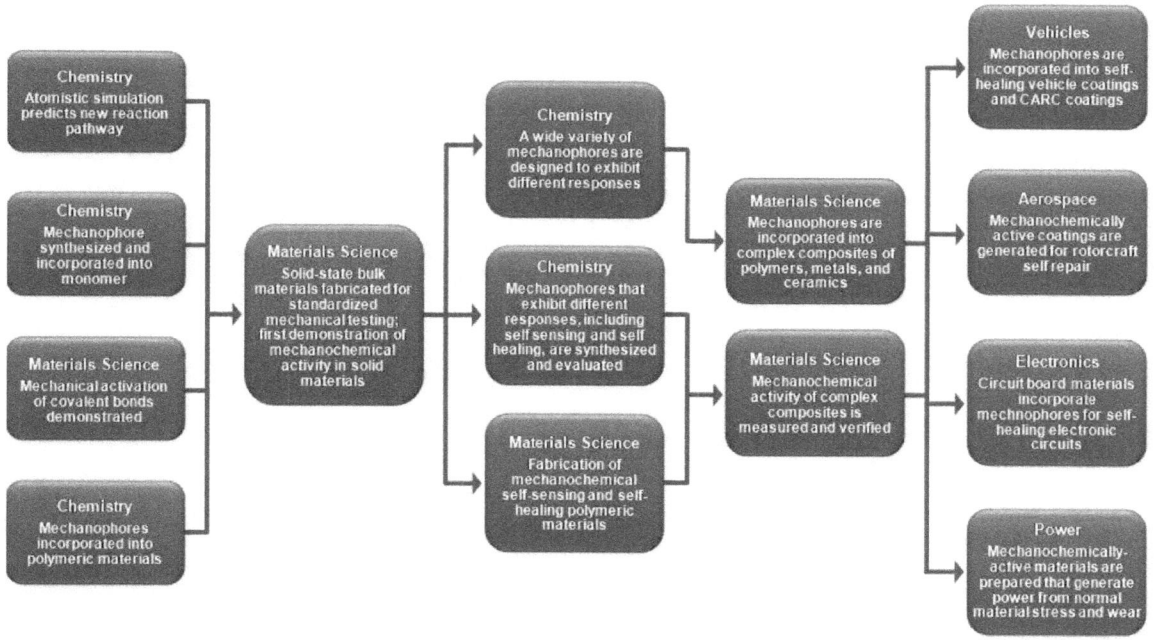

**Figure 5. Technology Convergence for Mechanochemical Transduction**

A report in the prestigious journal *Nature* commented on the first molecules developed through these collaborations, stating that "scientists from many disciplines will be

---

[43] M.J. Kryger, M.T. Ong, S.A. Odom, N.R. Sottos, S.R. White, T.J. Martinez, J.S. Moore, "Masked Cyanoacrylates Unveiled by Mechanical Force," *Journal of American Chemical Society,* 132 (2010), 4558–5799.

restricted only by their imaginations when it comes to finding ways of using mechanically-responsive polymers...."[37] Most importantly, the collaboration between these disciplines have opened the door to the conception of the new material designs. A close coordination between these disciplines in the future is essential to the success of this research, as each responsive molecule must be conceptually designed based on the desired material properties, chemically synthesized, and then carefully evaluated in a material system to determine whether it responds to a given force as designed.[41] The scientific opportunities provided by these ongoing, coordinated efforts could ultimately lead to i) applications in vehicles, including self-repairing armor, rubber, and coatings resistant to chemical agents, ii) aerospace applications, whereby aircraft and even satellites could be equipped with responsive coatings that self-repair when exposed to extreme temperatures or flight stresses, iii) circuit boards that can restore lost connections after physical or electronic interference, and iv) materials that generate energy from normal wear or stress, possibly even using energy from enemy attacks to power equipment.

# Quantum Information Science

Peter Reynolds, Marc Ulrich, and TR Govindan

In the 1980's and early 1990's, a small group of theoretical physicists and information scientists explored the idea of utilizing fundamental features of quantum systems for information processing.[44,45,46] The notion of a quantum bit of information (a qubit) was developed and a few algorithms proposed. In parallel, protocols for ultrasecure communications were proposed using the notion of a qubit and the potential capability of quantum systems to generate single and entangled photons.[47,48] Much of this research remained niche during this period, driven by novelty and constrained by the lack of experimental support that could point to physical implementations.

This picture changed dramatically in the mid-1990's by the publication of two algorithms that clearly showed the power of quantum systems to solve hard computational problems, thought beyond the capability of classical computers.[43,49] These algorithms along with the prior work on ultrasecure communications altered the foundations of information security. These developments were rapidly followed by a publication that showed that computational errors caused by inevitable environmental effects on a quantum system could be corrected.[50,51] Quantum error correction pointed to the feasibility of building quantum information processing systems and contending with the formidable task of isolating the quantum system from errors caused by the environment. During this period, atomic physicists recognized that the techniques they had developed to control and manipulate atomic systems (especially for atomic clocks) could be mapped to the notion of a quantum bit and logical operations on quantum bits.[52,53] The confluence of these

[44] Richard Feynman, "Simulating Physics with Computers," *International Journal of Theoretical Physics,* 21 (1982), 467–488.

[45] D. Deutsch, "Quantum Theory, the Church-Turing Principle and the Universal Quantum Computer," *Proceedings of the Royal Society of London, Series A,* 400 (1985), 97–117.

[46] D. Deutsch, "Quantum Computational Networks," *Proceedings of the Royal Society of London. Series A, Mathematical and Physical Sciences,* 1868 (1989), 73–90.

[47] C.H. Bennett, G. Brassard, "Quantum Cryptography: Public Key Distribution and Coin Tossing," (presented at the International IEEE Conference on Computers, Systems & Signal Processing, Bangalore, India, December 10–12, 1984).

[48] P.W. Shor, "Polynomial-Time Algorithms for Prime Factorization and Discrete Logarithms on a Quantum Computer, (presented at the 35th Annual Symposium on Foundations of Computer Science, November 20–22, 1994).

[49] L.K. Grover, "A Fast Quantum Mechanical Algorithm for Database Search," (proceedings of the 28th Annual ACM Symposium on the Theory of Computing, 1996).

[50] P.W. Shor, "Fault-Tolerant Quantum Computation," (presented at the 37th Annual Symposium Foundations of Computer Science, 1996).

[51] A.M. Steane, "Efficient Fault-Tolerant Quantum Computing," *Nature,* 399 (1999), 124–126.

[52] J. Cirac and P. Zoller, "Quantum Computation with Cold Trapped Ions," *Physical Review Letters,* 74 (1995), 4091–4094.

[53] C. Monroe, D.M. Meekhof, B.E. King, W.M. Itano, and D.J. Wineland, "Demonstration of a Fundamental Quantum Logic Gate," *Physical Review Letters,* 75 (1995) 4714.

events and convergence of research in physics, mathematics, and computer science led to conceptual ideas for a quantum computer.[54,55]

During the last decade, QIS research has largely focused, with much success, on exploring physical implementations of qubits and qubit operations and new applications of quantum information processing. Several physical systems are being explored for the suitability as qubits. Qubit operations have been demonstrated in several of these systems, including trapped ions, photons, cold atoms, superconductors, and semiconductors.[56] Materials and fabrication techniques for qubits have been improved to obtain better performance. Testbeds for quantum communications operating in telecommunication networks have been demonstrated.[57] Additional new algorithms have been discovered. The last few years have also seen a widening of the field beyond computing and communication. Parallel research in atomic systems combined with progress in QIS has led to the rapidly growing field of quantum simulations. Similarly, new techniques for sensing, imaging, and precision measurement with potential capabilities beyond those of classical systems are being explored.[58,59] The entire mapping of these, as well as additional, technological convergences is shown in Figure 6.

The current capability to control a few qubits, combined with developed implementations and new algorithms, has provided new insights into the role of quantum resources (superposition and entangled) in overcoming limitations of classical systems. These have led to a broad, unified, multidisciplinary view of the field of quantum information science with potential revolutionary impact on a broad range of applications. In all cases, capabilities that overcome the limitations of classical systems are envisaged and cover applications in computing, communications, simulation, sensing, imaging, and metrology. All these capabilities are core to the Army mission, with potential revolutionary impact on performing the mission.

---

[54] Nielsen MA, Chuang IL. *Quantum Computation and Quantum Information.* Cambridge University Press, Cambridge, UK, (2000).

[55] DiVincenzo DP. "The physical implementation of quantum computation," *Fortschritte der Physik,* 48 (2000), 771–783.

[56] T.D. Ladd, F. Jelezko, R. Laflamme, Y. Nakamura, C. Monroe, J.L. O'Brien, "Quantum Computers," *Nature,* 464 (2010), 45–53.

[57] SwissQuantum. "Swiss Quantum," SwissQuantum, available at <http://www.swissquantum.com>.

[58] Ignacio Cirac, Peter Zoller, "New Frontiers in Quantum Information with Atoms and Ions," *Physics Today,* 57 (2004), 38–44.

[59] T. Rosenband, D.B. Hume, P.O. Schmidt, C.W. Chou, A. Brusch, L. Lorini, W.H. Oskay, R.E. Drullinger, T.M. Fortier, J.E. Stalnaker, S.A. Diddams, W.C. Swann, N.R. Newbury, W.M. Itano, D.J. Wineland, J.C. Bergquist, "Frequency Ratio of Al+ and Hg+ Single-Ion Optical Clocks; Metrology at the 17th Decimal Place," *Science,* 319 (2008), 1808–1812.

# Convergence of S&T: Quantum Information Science

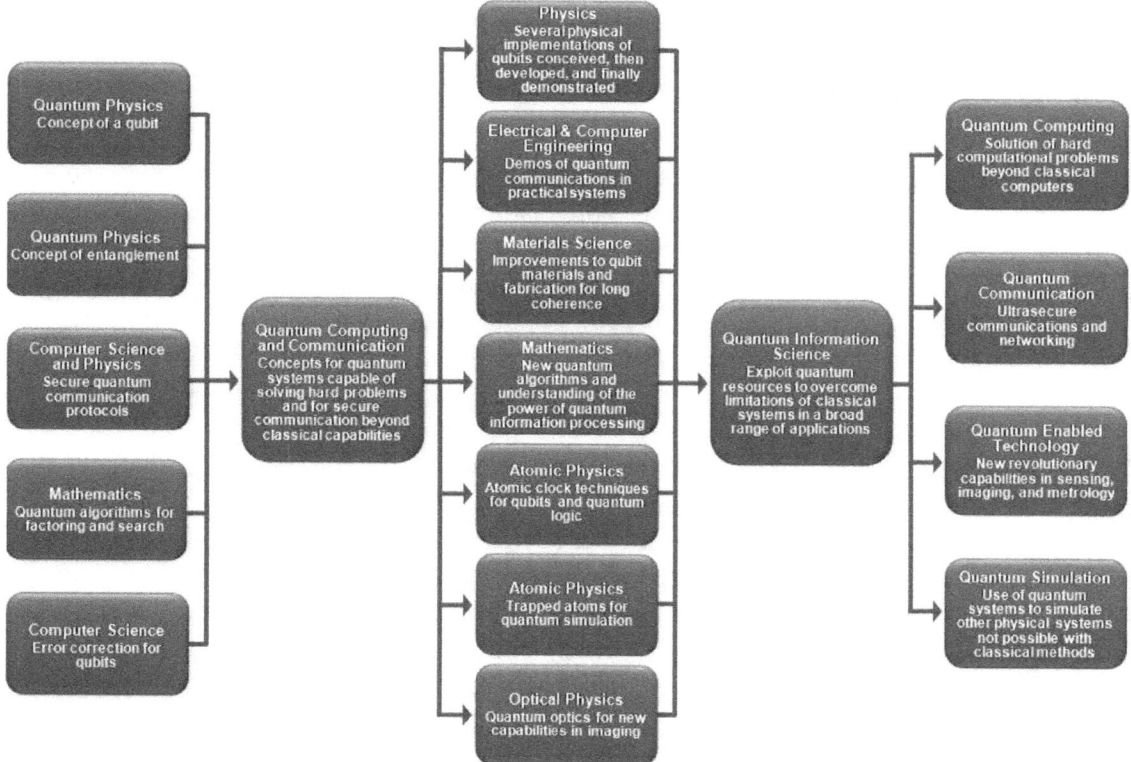

Figure 6. Technology Convergence for Quantum Information Science

# VII. The Army's Next Study—Comments and Recommendations

## *Forecasting Methodology*

The main focus of this paper has been on concepts of technological convergence and forecasting future developments in S&T. Forecasting is difficult due to the unpredictable nature of technology and often exponential technological advances. These often result from convergences. While it is difficult to predict when, where, and how S&T convergences will occur, it is possible to create an atmosphere within the Army's research planning and investment activities that enhances the probability that such convergences can be predicted.

Whether in industry or the military, the processes of research, development, and engineering are driven by mission requirement "pull" and S&T "push." Both must be incorporated in a dynamic manner in S&T forecasting. Within the Army, several TRADOC documents describe required Force Operating Capabilities, which are tied to the previously discussed warfighter outcomes.[7,8] The task is that of defining an end state for the Army. This end state is not an arbitrary point in time, but should be a continuum of near, mid-, and far-term capabilities responsive to the changing tactical and strategic demands of the Army's missions.

The methodology we recommend should produce a formal strategic roadmapping process that provides a timeline of three (near, mid, far) 5-year segments. The process can begin at either end (e.g., the S&T push or the mission pull); in an organization such as the Army, the latter is usually the starting point. The end state capabilities required to successfully prosecute the mission are first defined and generally based on the ultimate far-term requirements. The next step is to identify the actual operational capabilities, without regard to the current maturity of S&T, which would be required to achieve those end states. Simply put, what abilities must the warfighter have to accomplish his tasks? Following this, one identifies the science that enables operational capabilities by assessing current state-of-the-art developments and projecting likely advances given levels of investment. This process is iterative and requires a combination of independent subject matter experts (SMEs), Service laboratory S&T personnel, and warfighters.

The development of the initial strategic roadmap can be viewed in three, bidirectional steps: know why, know what, and know how. These steps correspond to the mission, the operational capabilities, and the S&T. A fourth step consisting of program planning; execution completes the process. During the roadmapping process, key S&T areas are defined and assessed, and likely temporal trajectories are binned according to 5-year segments. Some of the specific issues this process addresses are gating and risk factors, current investments both within and outside of the Army S&T community, opportunities to leverage such investments, key S&T performers, Army S&T infrastructure (both labs and people), the Army's role (e.g., lead, leverage, maintain cognizance) and potential for S&T convergences.

This formal roadmapping process directly overlays the S&T challenges to the mission (e.g., the push to the pull), can be viewed in either direction, and can be mapped to current S&T investments. The detailed analyses described in the previous paragraphs provide robust criteria with which to defend current investments, justify requests for additional funding, and (most important) provide a dynamic framework for forecasting S&T developments.

Perhaps, an additional check can be provided by proposed methodologies and use-case analyses to assess military benefits of S&T investments.[60,61]

## *Recommendations*

To achieve this dynamic framework for forecasting S&T developments, certain criteria must be met. As highlighted in the previous examples, our recommendations for efficiently and effectively developing a workable roadmap for such studies must include the following:

1. *A list of critical underlying sciences for a study should be grouped into clusters based on the list of capabilities from the first stage.* The clusters may be of closely related sciences, or they may be a mixture of rather different disciplines that might be synergistic. An example of the former could be chemistry, nanoscience, surface science, and material science. An example of the latter is the aforementioned study of nanoscience, biotechnology, information technology, and cognitive science.

2. *Each individual science in a cluster is first projected into the future.* Each science would be laid out on a timeline with significant likely advances noted at the projected times. The projections for each member of the cluster would then be plotted at the same scale.

3. *Fast-moving technologies should be studied together.* This can be done if the technologies are related. This is the case today for electronics, computers, and communications.

4. *The plots for each cluster are assessed to see where convergences may occur.* The convergences are assessed for their potential for something new and unexpected. At some point it becomes clear that the sciences are morphing into technologies with the possibility of conceiving prototype devices and components.

5. *The membership of the study committees should be balanced among the various disciplines or sub-disciplines involved.* The composition of the study committee can have a major effect on the results. Inevitably, the committee report will reflect in some measure the opinions, often strongly held, of the members. Care should be taken to avoid allowing any one viewpoint or personality to dominate through public identification of biases. Separate expert panels representing each cluster of

---

[60] Albert Sciarretta, Richard Chait, Joseph Mait, and Jordan Willcox, *A Methodology for Assessing the Military Benefits of Science and Technology Investments*, Defense & Technology Paper 55 (Washington, DC: Center for Technology and National Security Policy, National Defense University, September 2008).
[61] Albert Sciarretta, Richard Chait, Joseph Mait, and Jordan Willcox, *Assessing Military Benefits of S&T Investments Utilizing Use-Case Analyses*, Defense & Technology Paper (Washington, DC: Center for Technology and National Security policy, National Defense University, to be published).

science serves as an economical and effective approach to highlighting convergences.

6. *SMEs from the Army should be allowed to participate beyond liaison roles.* One possibility is to require that some number of SMEs, perhaps drawn from the Army's senior scientists (STs) or division chiefs in the laboratories, make presentations at the outset of the committee deliberations and be allowed to attend the meetings as observers and resource providers. (They should not, however, play any part in the committee's operations nor should they participate in the writing of the report.)

# VIII. Conclusion

This paper makes the case for approaches to be pursued when the Army conducts its next comprehensive S&T forecasting effort. A worthwhile exercise, forecasting can sharpen planning in the short term for aligning the research S&T programs with TRADOC's list of needed capabilities, selectively strengthening some areas and adjusting the research and development budget accordingly. Focusing upon human-centric and disciplines related to situational awareness, as well as the scientific underpinnings of new fields, provides a greater possibility to identify potential convergences that can lead to desired capabilities and outcomes. We have provided recommendations on how to prepare roadmaps and implement such efforts. Through clustering and mapping each scientific area, we believe such studies will result in convincing rationales, and that these rationales should engender a better understanding of the S&T program on the part of the Army leadership and Congress.

Finally, in the previous sections, we offered examples of clusters of sciences or technologies. These are meant to give a more detailed idea of what is meant by clusters and to further explain the concepts of convergence in technology forecasting. This paper did not carry out technology forecasting per se but rather provided examples that were intended to illustrate how the Army Science and Technology Executive should include the concept of technology convergence in a statement-of-task for the next comprehensive technology forecasting study conducted by the NRC or similar organizations.

# Appendix: Expanding on Human Warfighter Outcomes— Biobehavioral Response

In other relevant research, psychological hardiness has been shown to be related to a number of HPA (hypothalamus-pituitary-adrenal) axis stress response hormones, notably cortisol and b-endorphin.[62] These authors suggest that high hardy persons are less stress-reactive and that, while basal levels of HPA hormones were somewhat elevated, hardy persons are less volatile in reacting to stressors. In short, hardy individuals are more likely to remain emotionally stable during stressful situations.

In support of this interpretation is the observation that, for people generally, stress responsiveness is lowest at the circadian peak of glucocorticoid levels and highest at the circadian nadir. Looking at immune system functioning, significant differences between high and low hardy persons were identified, with the high hardy group showing more robust lymphocyte proliferation (T and B cells) in response to several infectious agents.[63]

Neurological correlates of hardy individuals and their stress response have been addressed in recent studies as well. A common glucocorticoid investigated in human stress interactions is cortisol. Increased cortisol levels are generally considered useful or adaptive for stress responding within limits, as cortisol affects multiple systems to increase energy and arousal allowing the organism to address acute threats more effectively. But, if these increases in cortisol and glucocorticoids should continue unabated, they can have a range of damaging effects on the organism, including on brain cells, particularly in the hippocampus.[64] Dehydroepiandrosterone (DHEA—an adrenal hormone) helps to modulate or restrain the effects of cortisol.[65] Summarizing results from several studies on DHEA and mental health and performance suggests that DHEA is a potentially important marker for stress resilience.[66] Some indirect support for this notion comes from two separate studies conducted by Morgan and colleagues, both with military groups. In the first, positive stress response (e.g., few dissociation symptoms) was associated with higher DHEA-S (DHEA-sulfate)/cortisol ratios for students in military survival school.[67] In the second study, positive stress response (few dissociation symptoms) was associated with higher hardiness (challenge) levels in Norwegian officer

---

[62] E.P. Zorrilla, R.J. DeRubeis, and E. Redei, "High Self-Esteem, Hardiness and Affective Stability are Associated with Higher Basal Pituitary Adrenal Hormone Levels," *Psychoneuroendocrinology*, 20 (1995), 591–601.

[63] C.L. Dolbier, R.R. Cocke, J.A. Leiferman, M.A. Steinhardt, et. al, "Differences in Immune Responses of High vs. Low Hardy Healthy Individuals," *Journal of Behavioral Medicine*, 24 (2001), 219–229.

[64] R. Sapolsky, "Stress and Plasticity in the Limbic System," *Neurochemical Research*, 28 (2003), 1735–1742.

[65] E.S. Browne, B.E. Wright, J.R. Porter and F. Svec, "Dehydroepiandrosterone: An Antiglucocorticoid Action in Mice," *American Journal of the Medical Sciences,* 303 (1992), 366–371.

[66] D.S. Charney, "Psychobiological Mechanisms of Resilience and Vulnerability: Implications for Successful Adaptation to Extreme Stress," *American Journal of Psychiatry,* 161 (2004), 195–2216.

[67] C. A. Morgan, S. Southwick, G. Hazlett, A. Rasmusson, G. Hoyt, Z. Zimolo and D. Charnery, "Relationships Among Plasma Dehydroepiandrosterone Sulfate and Cortisol Levels, Symptoms of Dissociation, and Objective Performance in Humans Exposed to Acute Stress," *Archives of General Psychiatry,* 61 (2004), 819–825.

cadets undergoing a stressful military exercise.[68] The same measure of dissociative symptoms was used in both studies. This suggests there may also be a correlation between hardiness and DHEA-S/cortisol ratio, but to date this possibility has not been fully explored.

---

[68] Eid and Morgan, "Dissociation, Hardiness, and Performance in Military Cadets Participating in Survival Training," *Military Medicine*, 171 (2006), 436–442.